WordPress:

A Step-by-Step Beginners' Guide to Build Your Own WordPress Website from Scratch

Table of Contents

Disclaimer

About the Author

John Slavio is a programmer who is passionate about the reach of the internet and the interaction of the internet with daily devices. He has automated several home devices to make them 'smart' and connect them to high speed internet. His passions involve computer security, iOT, hardware programming and blogging.

Introduction

A few years back, setting up your own blog was a geek's job. You needed to code your whole blog and then upload it online to your hosting. In a nutshell, it was really a pain to set up your blog back then. But today, it's as easy as blinking your eye. All you have to do is install a blogging platform on your hosting. Today, you have several platforms for building up your blog. Drupal, Blogger, WordPress, and many others will all enable you to go online and share your experiences and knowledge with the world. While building a blog, you need to design a look that aligns with your blog's topic; later, you need to add different pages to your blog to make it complete. Several blogging platforms make this task easy to do; however, setting up a blog is still a huge undertaking. Which blogging platform is the most efficient?

Out of all the blogging platforms, the best and most famous blogging platform is WordPress. This is because it makes setting up your blog simple and it provides you with the flexibility to make any kind of changes in your blog anytime you want. Moreover, with its plugins feature, you can add any kind of functionality to your blog. Do you want to set up your online store? Install the "Woocommerce" plugin. Do you want to make your blog SEO optimized? Install the "SEO by Yoast" plugin. All it takes is a few mouse clicks here and there to add such great features to your blog.

In this eBook, I am going to take you by the hand and show you how you can set up your own blog using WordPress starting from scratch. On our journey, I am going to discuss every single option that WordPress provides to make your blog more appealing and unique. Whether you are a beginner or an intermediate, with this eBook, you will surely have something new to learn. At the end, you will have enough knowledge to set up your own blog with ease.

Why Should You Select WordPress?

I always recommend everyone to go for WordPress. But why? *Why should I use WordPress and not any other blogging platform?*

Frankly speaking, there are not just one or two reasons to select WordPress. In fact, there are many.

- WordPress is an open source software. Which means it's free to download, use, and modify to make websites. You can download and install it from the official WordPress website. Moreover, today most of the hosting comes with a one-click install option to install WordPress automatically onto your blog or website. Being an open source software, WordPress is maintained by people who are always ready to help each other. So, if you run into any kind of problem, or need a patch or a new code to add some function to your website, all you have to do is just ask the WordPress community. Most of the questions get answered within 24 hours along with several solutions. Select the one you like and proceed.

- WordPress was the best blogging platform in the past, but now it has emerged as the best content management system (CMS), which can be used to make any kind of website. For example, you can use WordPress for...

  - Creating blogs

  - Creating online stores (e-commerce websites)

  - Creating a vBlog (video blog, also known as video collecting site)

  - Creating a portfolio website

  - Creating a review website (mostly used by affiliate marketers for reviewing products)

  - Creating a membership website

  - Creating a single page promotional website

  - Creating a knowledge base website (like wiki websites)

- Creating a forum for question and answers like quora.com
- Creating a gallery or photography website

And many more options. The possibilities are unlimited. You can create a website of your choice with WordPress; all you need is a topic and you are good to go.

- WordPress is used by almost 60% of the total websites on the Internet today. And even now, more and more people are joining with WordPress to build their websites. It is because WordPress is really easy to use and learn as compared to any other blogging platform. Most of the people who use WordPress have ZERO knowledge of coding and programming; as a matter of fact, most of the people who opt for WordPress have no knowledge of websites at all. So how are they able to still use WordPress? WordPress provides them with easy-to-use flexibility in building their website. WordPress has more than 2600 themes to select from for your website topic and more than 31,000 plugins which are available to you to add different functionality to your website with just a few clicks. It is not only easy to find the theme of your choice for your blog or website, it is fairly easy to customize it, too. In fact, every theme comes with its own customization panel which makes it simple to adjust the theme setting like header, colors, footer, etc., as per your need.

- WordPress is the most SEO friendly platform that you can have for your website today. Currently, SEO depends not only on the content that you write on your website but also on various other factors like the overall user experience, ease of use, ease of accessing navigation panel, having relevant pages, website speed, and image optimization. With WordPress, you don't need to worry for any of these concerns as WordPress is written with high-quality codes which cover all of these parameters and makes smoothly enables the user to optimize his website for search engines. According to Google, WordPress is 99% SEO friendly, which the highest rating over every other blogging platform.

- With other blogging platforms, most of the time you need to add codes manually to add functionality—and you will have to keep an eye on your codes so that you can stay updated. With WordPress, it's a snap to manage your website and the codes that you add in as plugins. WordPress gives a notification in its upper panel from where a user can update all the installed plugins and themes to the latest version with just a few clicks. All you have to do is select the plugin to update, click "update," and you are done! Easy, isn't it?

- WordPress is among the safest blogging or website-building platforms in the world. Powering almost 60% of the total websites on the Internet, it was developed keeping safety and security in mind. WordPress constantly releases regular updates in order to minimize the chances of your website getting hacked. Moreover, using certain plugins like "WP Security Firewall" will enhance your website's security further.

- WordPress is multimedia friendly. You can effortlessly insert almost any kind of media such as videos, images, and audio clips in your website using WordPress's rich text editor, "WYSIWYG."

- At some point in time, you will feel a need to integrate your website with different services, like email marketing services, or content creating services. With WordPress, you can easily integrate your website with almost any high-quality service on the Internet. You don't need to add manual codes for any service to work simultaneously with your website. WordPress handles it all, as most websites today have a one-click option to integrate any WordPress blog with them.

So as you can see, selecting WordPress is probably the best time investment that you will ever do when building your website. When selecting WordPress as a blogging platform, most beginners often get confused between WordPress.com and WordPress.org. Thus, before moving on, it is important to understand the difference between the two and to know which one we will be using in this eBook.

WordPress.org	WordPress.com
With WordPress.org, you will host your own website or blog. You will find the open source WordPress software at WordPress.org that you can download and upload on your hosting to install and use. Moreover, most of the hosting companies today offer a simple one-click install for installing WordPress on your website.	If you are looking for a free se don't need to find a host for y WordPress.com is best for you not have to pay for hosting an worry about downloading and software. WordPress takes cai here.
With WordPress.org, you will have to purchase a domain name and hosting where you can set up your own blog or website.	With WordPress.com you will domain ending with wordpres instance, if you select your we YOUREXAMPLE, then your blo – http://yourexample.wordpr Although WordPress.com allo purchase a domain name and .wordpress.com website, still .wordpress.com will be taken website address for your blog
WordPress.org gives total control of your website. You can upload custom codes, plugins, themes, and modify the PHP code	WordPress.com limits the con of your website. You cannot u of multimedia to your website

So, if you need full control of your website, then you will need to go for WordPress.org instead of WordPress.com. As this eBook is about building your website from scratch, I will be using WordPress.org in the rest of the eBook to help you further. In the next chapter, we will start

working on building a blog with WordPress.org and I will show you all

the prerequisites that you need to build a WordPress powered blog or

website.

Prerequisites for Creating a WordPress Blog

Okay, now we'll focus on creating a self-hosted WordPress blog or a website by using WordPress.org. Below is a list of what you will need in order to get started with building up your WordPress blog.

- A Domain Name

- Hosting

Both prerequisites are mandatory and the only investment that you will need in order to create a WordPress blog. Now, I'm going to explain these prerequisites, along with everything that you need to know about them and where you can buy them. So, let's start!

A Domain Name

Domain is the first thing that you will need in order to build a website or blog with WordPress. A domain name is what you type in the address bar of your Internet browser to visit a website—for example, Google.com is the domain name you use for visiting Google. A domain name allows anyone to access your website from anywhere in the world. If you have registered a domain already, then you're good. If not, then I am going to tell you where to purchase it later in this chapter.

WHOIS Guard (Optional)

When purchasing a domain, you get an option to buy WHOIS guard. Although it is optional, but I strongly recommend you buy it. The reason is because you provide your personal information like your name, address, and phone number when you purchase a domain name. This information is available publicly for everyone to see. Thus every day, domain owners' information is used by spammers to send spam. WHOIS Guard prevents such activities from happening by protecting your privacy and making your information unavailable for public viewing. That's why it's absolutely essential to never go online with any of your websites without a WHOIS guard.

Hosting

After purchasing your domain name, the next step is purchasing a hosting for your website. So, what exactly is hosting? Let's try to understand it with an example. Let's take a person as an analogy for a domain. Now that person keeps all his stuff in his house, and he stays in his house. So, if that person is a domain, then this house is a hosting. That is, a hosting is a storage facility where your website sits. For the website to appear online, it needs a hosting, where all the stuff required for its working (your multimedia, logo, favicon, etc.) can be kept.

As your website or blog will be running on WordPress, it is highly recommended to go with a hosting provider which is WordPress friendly. When purchasing a domain, you will get a lot of options to select from, such as WordPress hosting, Linux hosting, and Windows Hosting. From all of these, Linux hosting is the best and most recommended whenever you are building a website. Inside Linux hosting, every hosting provider has their different plans according to the bandwidth and storage they provide. Though it is not required, I recommend selecting a plan with unlimited bandwidth. Storage at the time of starting a website doesn't matter and you can upgrade your hosting plan whenever you like for more storage. Bandwidth is generally the amount of data transferred through the website at one time. The whole gamut of webhosting works on this parameter. Bandwidth affects your website speed and availability. If your bandwidth is less, your website speed may be slow and will sometimes force your website to show an error saying, "Bandwidth Limit Exceeded." Thus, in order to ensure proper working of your website at its maximum speed, I advise that you buy a Linux hosting plan with unlimited bandwidth.

Where to Buy Domain & Hosting?

When buying a domain and hosting, it's best to buy them from the same provider. This way, it'll be easy for you to link them together. If you buy a domain from a domain provider and hosting from some other provider, then you will have to follow a process of linking your domain to your hosting. The process is simple but for the sake for beginners, I recommend buying both from the same provider. Below is a list of top domain and hosting providers that I recommend.

GoDaddy.com – This is by far the best and most affordable domain and hosting provider in my opinion. I have been using Go Daddy since 2012 for all my websites and have never faced any problems with them. When you are making a new purchase, Go Daddy turns out to be most affordable. Several Go Daddy discount coupons make it easy to grab both a domain name and hosting for as low as $20 for the first year. The renew fee of Go Daddy is higher than other service providers like Bluehost.

How to get both hosting and domain for $20 by Go Daddy

When on Go Daddy's website, go to hosting and select 'Economy Linux Hosting' for one year; this way, you will be offered a free .COM domain by Go Daddy with your purchase. You will only need to buy an extra WHOIS guard, which is optional, as I discussed before. Go Daddy always has several coupons active on 'Economy Linux Hosting' which makes it possible for grabbing hosting, domain, and WHOIS guard for $20. The best website for looking for such discount coupons is "Retail Me Not."

Bluehost – Just like Go Daddy, Bluehost has never let me down with its domain and hosting service. Although Bluehost is as good as Go Daddy, the amount that Bluehost charges for your purchase and for the renewing of your domain and hosting is much higher than Go Daddy.

Namecheap – This is another service provider that I recommend for those who are looking for an affordable yet competitive hosting service that also provides a domain name. Unlike Go Daddy, you will have to buy both domain name and hosting on Namecheap. The WHOIS guard on Namecheap comes free with every domain that you purchase. The renew fee of Namecheap is the least among the three service providers and I have never faced any problems with them before. Few of my sites are still running on Namecheap which I started before I moved to Go Daddy.

My personal recommendation for hosting and domain is Go Daddy.

Finally, Go Daddy is unbeatable in support, in my opinion. They offer excellent phone support and are always ready to help you out with any problem related to your purchase. You are still free to buy from any of the three services that I told you or from any other service provider that you know is reliable.

Getting Started with WordPress

Okay, by now I assume that you have purchased a domain name and hosting for your website. Your next step is to install WordPress on your website. So, let's start!

Installing WordPress

As we bought Linux hosting, I want to tell you that every Linux hosting comes with a control panel, which is also known as cPanel. Installing WordPress is easy with cPanel as the process is automated. The first thing you need to do after purchasing a domain and hosting is to link them up. Now, if you bought them from the same service provider, then linking them is really easy. In fact, you are helped in linking them by your service provider. If your hosting plan is limited to one website, then some hosting providers already link your purchased domain and hosting for you. I am using Go Daddy right now and I will explain how to install WordPress using cPanel focusing on Go Daddy. But know that the process is still the same for all the service providers. Here are the steps:

- First, link your domain to your hosting. Most of the service providers make it easy to do it.

- Now, the next step is to access your cPanel. There are two ways to do this.

  - In Go Daddy – Log into your account. On the next screen, you will see *DOMAINS, WEB HOSTING, WORKSPACE EMAIL.* Select "web hosting" and click "manage." It will take you to your cPanel login page. Enter your login details and you will be inside your website cPanel.

  - In Bluehost – go to http://my.bluehost.com and from the selection tab of "hosting login" and "webmail login," click "hosting login" and enter your login details in the form just below that. After entering the details, you will be taken to your website cPanel.

- Another way to enter your cPanel is by writing "cPanel" at the end of your website URL. For example, if your website name is http://www.yourwebsite.com, then write the URL in as http://www.yourwebsite.com/cpanel. This will take you directly to the cPanel login page. Enter the details there and you will be inside your cPanel in few seconds. I use this method to log in to my cPanel.

- Now, after logging into your cPanel, the next step is to find the install WordPress button. Go Daddy has updated its cPanel user interface and now the "Install WordPress" button comes at the top inside "Your Building Tools." If you are using Bluehost, just scroll down to "websites." There you will see the "install WordPress" button. From that button, you will be able to install the WordPress platform on your website with a few clicks. Just select the domain that you want to install the WordPress on, and select a WordPress username and password for your website. You can enter other information like the website title and tagline, but right now, but I prefer you leave them as I am going to explain the WordPress settings tab later in this eBook. That will let you change everything from the title and tagline to the way your website link appears on the Internet.

Logging into Your WordPress Website

Congratulations! You have successfully installed WordPress on your website and now it's time to log in to your WordPress blog or website for the first time. To log in, write the URL of your website followed by "wp-admin." So, if your website name is http://www.yourwebsite.com, then write it as http://www.yourwebsite.com/wp-admin.

You will now see the WordPress login page. Enter the username and password that you created while installing the

WordPress before. After logging in to your WordPress website for the first time, you will be welcomed by the WordPress Dashboard.

The WordPress Dashboard

When you log into your website for the first time, you will be taken to the WordPress dashboard. So, what is this WordPress dashboard and why are you taken here after you log in? WordPress dashboard is the place from where you control, modify, and manage your website. Everything you do on your dashboard is visible to you until you save the changes, so it's a kind of back-end or behind the scenes place for your website. Keep reading and you will see how easy it is to use this WordPress dashboard for doing anything with your website.

As soon as you log into your WordPress website, you'll see a top "Welcome to WordPress" section with some quick links to help you get started. You can always dismiss this box using the link in the top right corner.

The next section on your dashboard home screen is "at a glance." In this part of the home screen, you will see the number of posts and pages, which is the total content available on your website. Also, you will see the number of comments that are there on your website, giving you an idea of the discussion that is going on. At the end of this section, you will see the WordPress version that you are running, along with the name of the theme that your blog or website is using right now.

The next section just below the "at a glance section" is the "activity" section. Here you can see the information like recently published posts, comments, and pingbacks. From this section itself, you can approve the pending comments or can delete them to spam; other features like "unapproved" and "trash" are also available.

In the second column, at the top, you will see the "quick draft" section. This section is different from other sections available on the dashboard home screen. In this section, you can create quick drafts—

i.e., ideas for a new post that you want to write or perhaps some information related to the pages on your website. You can save a draft directly from this section of the home screen.

The next and the last section of the dashboard home screen is "WordPress News." From this section, you can read all the official news related to WordPress.

At the left side of the dashboard is a navigation menu. This navigation menu makes it possible for the admin to make posts; upload multimedia content like audio, video, and images; make pages; add themes and plugins; and create widgets. We'll explore these possibilities later on in this eBook.

WordPress Admin Bar

WordPress admin bar is another great feature of WordPress platform. It allows the user to easily edit and manage the site directly from the site without the need to jump back to WordPress dashboard every time. The WordPress admin bar is available at the top of the front end of the website. The "front end" is the part of the website that your visitors see—that is, the actual website—while the "back end" is the part of the website which only the admin can see and access—that is, the WordPress Dashboard. It provides some handy shortcuts to access parts of your WordPress installation without having to find them in the left-hand navigation menu. I am now going to explain this admin bar a little more so you know how it actually works.

After you log in to your WordPress blog or website, you will see a blackish ribbon at the top of your website. This ribbon is your admin bar. On the extreme left side, you will see a WordPress logo. If you will swipe your cursor over the logo, it will display some links to WordPress related information. We are not interested in all those links except the "Support Forum" link. If you ever run into any problem or want to ask something, just click the Support Forum link from the WordPress logo and you are good to go! I have already told you how helpful this support forum is in the eBook before.

Next, after the WordPress logo, you will see the name of your website. If you will take your cursor to the name, you will see the option to "Visit Site." From this link, you can see the front end of the website if you are currently on the back end of your website, and vice versa. Next to your website name is the comment section. This section gives you a quick overview of the comments on your blog or website. This section will have a number against it, telling you the number of comments which requires your attention, including the new comments made on your blog. The next and final section in this admin panel is "+New." If you will roll your cursor over this section, you will see a lot of links to the post, page, media, and user. From this section, you can add these items easily and quickly. If you open a front end page or post of your website, then you will see a new section in the admin panel called "edit." With this section, you can quickly go to edit your page or post.

These are the sections that you need to know about in the admin bar. Anything else that exists in the admin bar because of a new plugin, for example, works in the same way. In the next chapter of this eBook, we will discuss the WordPress Setting Section thoroughly. From here, your WordPress blog will actually start to build up.

WordPress Settings Section

I told you already that WordPress provides a lot of options for customization and adding functions to your websites. For a start, you can begin customizing your website from the WordPress setting sections. Moreover, I told you to not pay any attention to various fields like Title and Tagline while installing the WordPress. We will change everything here in the WordPress settings section. To access this settings section, all you need to do is go to your WordPress dashboard and locate "Settings" in the left navigation menu. When you go to the settings, you will see various WordPress settings –

- General
- Writing
- Reading

- Discussion

- Media

- Permalinks

I am going to explain each of these settings submenus one by one to you so that you will quickly understand them.

General WordPress Settings Section

You can access this section by clicking the "General" submenu in settings. Let's now discuss the various settings that we can change.

The first option in this general settings section is "Title and Tagline." Here, you will need to enter your website's title and tagline. For example, my new website is weightlosswellbeing.com; thus, I selected "Weight Loss Wellbeing" as the title and "Lose weight healthily" as my tagline.

Always make sure that the title and tagline match your website name and topic. This is because the title and tagline you select will be visible for your website in search engines like Google, Bing, and Yahoo.

There is a default tagline that WordPress puts on every website which is – "Just Another WordPress Website." Remember to change it to the one you want to keep for your website.

The next option in general settings is "WordPress Address." This is the URL of the website where you installed WordPress. This option is followed by "Site Address." By default, it is the same as the WordPress Address, but if you want your homepage address to be different from the WordPress Address, then you can enter the new homepage address here. Since we are developing a blog from scratch, I prefer you to leave this setting as it is.

The next option in general settings is "Email Address." This is the option where you need to put in your email address where you want to receive all the notifications related to your website. I prefer you to enter an email address that you mostly use. Also, I recommend that you sync that email over your smartphone so that you can receive all the notification emails instantly and can attend to them if something requires urgent attention.

The next option is "Membership." With this option turned on, you can allow anyone to register on your website, and every new registration is given a default role as a subscriber by WordPress. You can change this to "contributor," "editor," etc., from the "New User Default Role" option, which is just below the "Membership" option. For this, I suggest that you stick to the default role of the subscriber for new users—unless you want to give them editing or writing access on your website.

The next option in general settings section is "Timezone" where you can select your country's time zone. This is provided in order to sync your blog posts timing with your local time, as the time you make a blog post is always visible to the readers on your website unless you turn it off. The next three areas, "Date Format," "Time Format," and "Week Starts On," allow you to customize your date and time settings.

In case you want to change your website's language from English to any other language, you can do that from the last option available in the general settings section called "Site Language." However, to make your website available for the readers worldwide, I advise you to keep your blog's language as English. Once you are done with modifying general settings, always click "save changes" to save your modifications.

WordPress Writing Settings Section

You can access this setting section by going to settings and clicking "Writing." These are the settings which apply to the new content (posts) that you write and publish on your website. It has two sections inside it; the top section controls the editing and publishing of content from within the WordPress dashboard, while the bottom section controls publishing content from external sources. The first section has four options which are formatting, default post category, default post format, and default link category. I prefer you to start from the default post category as the formatting option doesn't require any attention. By default, WordPress keeps all the posts under the category named "Uncategorized." You can create categories in the post and can later select any one category as a default post category in the writing section. I am going to discuss category section briefly in the next chapter of this eBook. The next option is the default post format; it is selected by default as "Standard." But you can change it to any other option from the drop-down menu as per your website type. The next option is the default link category; it is set as "Blogroll" by WordPress and no other option is available to select.

The next section in the writing settings is "Post via email." Here you will be able to write content on your website with the help of your email. All you have to do is send an email to your website with the post content. To use this, you'll need to set up a secret e-mail account with a POP3 access, and any mail received at this address will be posted. For this reason, it's a good idea to keep this address secret.

The last option in writing settings is "updated services."

Whenever you make a new post on your WordPress blog, WordPress

will automatically notify the services listed in update services about

the new content. By default, there is just one service added, but you

can add more services which you can find with a simple Google search.

Once you are done modifying your writing settings, always click

"save changes" to save your modifications.

WordPress Reading Settings Section

The previous settings modified how the content is written and

published, whereas the reading settings change the way your content

is visible to your readers. You can access the reading settings by going

to Settings and clicking "Reading." The first option that you will see on

your reading settings section is "Front page displays." In this option,

you can select what should appear on your homepage. If you select

"the latest posts," your most recently published post will be displayed

on your homepage followed by the other posts in the same pattern.

However, if you select "a static page," then you will have to select two pages. One will act as your homepage, and one will be where all your latest posts will be visible. You can select the pages from the dropdown menu, but in order to have any pages in the drop-down menu, you need to create a few first.

The next option on the reading settings is "Blog pages show at most 5 posts." The number five is kept as default by WordPress. What it means is that on a single page, a user will see at most five of your blog posts at a time. If you have more than five posts on your blog, then he will see "previous posts" and "latest posts" button at the end of the page to navigate for more of your posts.

The next option is where you can control the display of your content in RSS feeds, including the number of recent items syndication feeds show and whether to show full text or a summary.

The last option available is "search engine visibility." If you don't want the search engines to know about your website and display it in search results, then check the option which says "Discourage search engines from indexing this site." However, I recommend you to let the search engines know of your website in order to gain traffic.

Always remember to click "save changes" after making any changes in reading section for them to take effect.

WordPress Discussion Settings Section

The discussion settings section provides several options to the admin for controlling and managing the comments made on his blog along with managing the links that are included by many people inside their comments. You can access discussion section by going to settings and clicking "Discussion."

Under the discussion settings, the first setting available is "default article settings." Here an admin is provided with three options. The first setting deals with links you make to other blogs. The second deals with pingbacks and trackbacks, or links back to your blog. The third setting is about your blog posts. This setting allows people to make comments on your blog posts. If you uncheck this setting, then the people will no longer be able to make comments on your blog posts.

The next setting available in the discussion section is "other comment settings." It has a total of five options available for the admin. These settings act as a guideline for how people can post their comment on your website. The first setting makes it mandatory for the reader to enter his name and email in order to make a comment, while the second option makes it compulsory for the reader to be a registered user on your website to make a comment on your website. Besides these two settings, I advise you not to change any other setting in this section.

The next section in the discussion settings is "email me whenever." You can understand this section just by reading; it doesn't require any kind of explanation. The next section is "before a comment appears section." This section deals with how comments are published. Here you can choose if an administrator must always approve comments or to publish automatically if the comment author had previously posted a comment.

The next section is "comment moderation." In this section, you can add when to hold a comment for approval. By default, WordPress holds a comment for approval if it has more than two links. You can change it according to your wish. Just below this setting, you will see a box section. In this box, you can write the names, words, IPs, and URLs for which you want to hold a comment for approval before it is visible on your website. The next section, the "comment blacklist" section, works in the same way to help the admin in managing comments on his website. The next section is the avatar section. An avatar is nothing but a general profile picture that you have assigned to your account for making comments. In the same manner, you can assign a picture to the people who are making comments on your website. You can select a default avatar for all the people or can filter them by rating. The avatar you selected for the people will display only if the people who are making a comment on your website do not have an avatar of their own.

Finally, remember to click "Save Changes" for your modification to take effect.

WordPress Media Settings Section

The media section allows you to make changes to the images that you upload to your website. You can access the media section by going to settings and clicking "Media." The first setting that you will see in Media is "Image Sizes." Under this setting, you can set the default size for different image types like thumbnail images and small-sized, medium-sized, and large-sized images.

Just below this section, you will see the "Uploading Files" section. In this section, you can change the default path of all the multimedia that you upload on your website. By default, the path is – yourwebsite.com/wp-uploads. You can change the URL to suit your needs; however, I recommend that you let it be as it is. And don't forget—if you make any changes in the "Image Sizes" section, be sure you click the "Save Changes" button for your modifications to take effect.

WordPress Permalinks Settings Section

Permalinks are the permanent URLs for your posts and pages on your WordPress blog. You can access this section by going to settings and clicking "Permalinks." Under this section, the only thing that requires our attention is the "Common Settings" section. By default, WordPress selects the plain type permalinks for all your posts and pages. This permalink type is not SEO friendly and thus I suggest that you change it to either "Month and Name" or "Post name." Changing your permalinks in this manner makes them SEO friendly. Remember to save all your modifications before leaving the permalink section.

So, that's it! I have covered the WordPress settings section thoroughly with you. By now, your WordPress blog is almost set up. Now, we need to make it look more appealing and engaging to your readers before we go on to add some content in there. In the next chapter, I am going to discuss Themes with you in detail.

WordPress Themes

You've set up your blog and changed few settings from the WordPress Settings section. It's now time to get a theme for your website. So what is a theme? In WordPress, the theme is what provides the styling and look of your website. A WordPress theme controls how your website is presented to the readers. By default, WordPress installs a basic theme on all the blogs and websites. With WordPress, you can change the theme and the entire look of your website in few clicks, including the styling, formatting, user experience, and navigation without affecting your website content. Today, themes do a lot more than just styling and adding to the user experience. Now they are coded to control the functionality of your website, as well. For example, if you have a responsive theme on your website, then it will add a functionality to your website to change its look according to the device screen size. Themes are just a part of WordPress which makes it the most powerful and beloved blogging platform on the Internet today.

Selecting a theme is always linked to the topic of your website. Always have a rough outline of how you are going to make content available on your website. This step is essential in order to select a theme for your website. Today, you can find themes based on different categories and even on specific niche topics. Thus, a rough outline of your website content in mind can really help in selecting a better theme for your website. You can access the themes from the Appearance panel from the left navigation menu on the WordPress dashboard home screen.

Free Themes and Paid Themes

Most of the themes that you will see via themes section on your website are free to use; all you have to do is select a theme, and click "install." Later you can see its live preview before you activate it to set it as your current theme for your website. However, along with free themes, WordPress also has thousands of paid themes available. So, what is the difference between free and paid themes? Free themes can be found directly from your WordPress blog and can be installed; however, you will not be provided with any help related to patches that you may want to have on that theme. Also, the codes used in most of the themes are not that reliable. Premium themes, on the other hand, are made by high-quality codes which make them reliable. The developer of the theme will work along with you on your purchased theme in case you run into any problem or want to have a new code inserted in it. Premium themes are more flexible in customizing your website and are responsive, too (I told you what responsive means before). Besides, they are faster than the free themes and provides a better user experience, hence making your website even more SEO friendly.

How to Install Themes?

Okay, so how do you find and install a WordPress theme on your website? There are two methods to do this:

- The first method is to go to your WordPress dashboard and find "Appearance" in the left navigation menu and select themes. Now, click "Add New." You will now see a lot of themes which you can select and install on your website. Now, go to "Feature Filter" and select the following settings for finding the themes for your website or blog. For "Layout", choose "Two Columns." For "Subject", pick "Blog," and for "Features",   select the one you like and click "Apply filters." You can select more than one option in a subject in case your website is based on some other subject along with blogging. The same is true for the layout and features. After finding a theme, click on "Preview" to see how it looks like, and if you are satisfied with it, then click "Install." WordPress will now install the theme and will ask you if you want to see a live preview of the theme or want to apply the theme to your website or if you want to return to the theme page.

- Another way is to manually upload a WordPress theme. This method is used when you are trying to install a theme that was not available via the themes section on your WordPress blog or when you purchased a premium theme. All you need is your theme in .ZIP format. Go to Appearance in the left navigation menu of your WordPress dashboard and select themes. Now click on "Add New." On the next page, click "Upload Theme." Now select the theme in .ZIP format and click "Install Now." WordPress will now upload your theme and will install it on your website. Later you will be asked if you want to see a live preview of the theme or want to return to the theme page like it asked in the previous method.

If you want to go for a premium theme, then the best place to buy premium themes for your website is from themeforest.com. Themeforest has several premium themes with amazing features and customization options that you can purchase and use straightaway. However, if you are just looking for a blogging theme, then I advise you to go for themes powered by Genesis framework. All of the top blogs made with WordPress are powered by Genesis framework today. Genesis framework is created by StudioPress. It makes your website faster and more SEO friendly than before. However, when you purchase Genesis framework, you will also have to purchase a Genesis framework supported theme. You can find these themes at StudioPress itself.

WordPress Plugins

WordPress is an open source software, and thus it is supported by several third-party tools which, when uploaded to your website, can extend the functionality of WordPress so that it can now perform certain actions which it was not able to perform before. Just like themes, some plugins are free and same are paid. Everything is easy to install.

Unlike themes, you will find most of the plugins are available for free on WordPress. More than 30,000 plugins are available on WordPress for free. You can pay for some premium plugins if you want to add some unique features like a squeeze page (thrive landing pages) or want to make a membership website (optimizepress), for example.

However, the main difference between free and paid plugins is the support. Most of the time, plugins play nicely with the core of WordPress and with other plugins, but sometimes a plugin's code will get in the way of another plugin, thus causing compatibility issues.

With a paid plugin, you get support from the developer who will help

you out anytime you run into a problem due to the plugin, which is not

possible with the free plugins.

How to Install Plugins?

You can install plugins in two ways, just like you installed themes in the

previous chapter. Both methods are simple, if you follow these steps:

- The first method is to search for the plugin and install it from
  within your WordPress website. To do so, from your WordPress
  dashboard home screen, go to left side navigation menu and
  find the plugin. Now you will see three links; click "Add New."
  On the next page, you can search for a plugin and click "Install
  Now" to install it on your WordPress blog. You can access all
  the installed plugins from the "Plugins – Installed Plugins"
  section on your website.

- Another way to install plugins is to manually upload them in .ZIP format. From the "Add New" page in the previous step, click "Upload Plugin" and select the .ZIP plugin file that you want to upload, and click "Install Now."

Must Have Plugins

Below is a list of all the plugins that I recommend you to install and use on your website –

- Clef plugin for secure login

- SEO by Yoast for SEO purposes

- Contact Form 7 for creating a contact page

- Favicon XT-Manager for uploading a favicon

- Header and Footer for adding HTML codes to header and footer of your website

- WP Login Limit to limit unauthorized access to your website

- WP Super Cache for increasing your website's speed

In the next chapter of the eBook, I am going to explain you Widgets & Menus which is the last section we will discuss before moving on to creating content for your blog.

WordPress Widgets & Menus

Widgets and Menus present interesting capabilities on WordPress. Starting with the widgets, they are added to various sidebars in your WordPress theme (left sidebar or right bar or footer sidebar) to add more content to your website or to increase its features. A few examples of widgets include a search bar, your website social profiles, your newsletter subscription form, your "About" page snippet, and featured posts.

Widgets do add more features to your website, and just like plugins, they do not need any kind of coding knowledge. All you have to do is just drag the widget that you want to use and drop it in the sidebar that you want it to show up in. Keep in mind that the number of sidebars that you can have on your website is decided by the theme that you are using. In WordPress, there are two kinds of widgets that you can find: the widgets which come by default and the ones which appear in the widget section as a result of some plugin installation or theme installation.

How to Add Widgets?

Adding widgets is an easy drag-and-drop mechanism, like I said earlier. Just follow the bullet points below and you will be able to add widgets to your website in no time.

- From your WordPress dashboard home screen, in your left navigation panel, go to "Appearance" and then click "Widgets."

- Now, you will see the widgets section on your screen. After the left navigation panel, you will see various widgets that are available by default and other widgets which are there as a result of a plugin or theme installation. To the right side, you will see various sidebars that your theme has where you can add the widgets.

- Now, just click and drag the widget from the left and drop it in the sidebar on the right. You can also arrange the widgets in a sidebar according to your needs by dragging and dropping them up or down.

Now, let us move to the Menu section. Menus are created for providing your reader with easy navigation on your website. Depending on your theme, you can have a number of menus, like the primary menu, secondary menu, footer menu, etc.

There is no limit on how many menus you can create in WordPress; however, how many you can use depends on the theme that you are using.

How to Create Menus?

Below are detailed instructions that you can follow to create a menu easily for your WordPress blog.

- From your WordPress dashboard home screen, go to "Appearance" and click "Menus" in the left navigation panel.

- Now you will see a menu editor page on your screen. Enter a name for your menu and click 'Create Menu."

- Now your menu will be created. You can now add pages, posts, categories, and even custom links from the options available on the left side of your newly created menu by clicking "Add to Menu."

- You can arrange the menu items just like you arranged the widgets—by dragging and dropping. After you have created a menu, click "Save Menu."

- You now have your navigation menu ready. Just below your newly created menu, you will see an option called "Theme Location." You will be able to select a location for your menu from here. After selecting a location, click "Save Menu" again to save the changes.

Creating Content

So your WordPress blog is all set up now, but no blog or website is called complete unless it has some content on it. Creating content is really easy in WordPress. However, you have two options in WordPress for creating your content, namely "Posts" and "Pages." Let's start by understanding the basic difference in between these two.

Difference between Posts & Pages

Posts and pages can be accessed from the left navigation menu on your WordPress dashboard. Posts are used for making general articles and updates on your website based on the topics that your website covers. Pages, on the other hand, are used for making static content, which does not require another article for an update. For example, an About Me page or a Contact page are static pages. Moreover, in posts, the articles you write can be organized in a structural manner by creating categories, whereas there is no such option available for pages. Pages are also not included in RSS feed and never display the time of publishing, unlike posts.

So, posts are used for creating blog content and pages are required for creating static content which doesn't require often updates.

How to Create Posts & Pages?

Creating posts and pages on your WordPress blog work in the same way. So, I am just going to discuss both of these options together in this section. Let's start by learning how to access the pages and posts in WordPress to create new content.

- To access posts, from the left navigation menu on your WordPress dashboard, click "Posts." On the next page, click "Add New."

- To access pages, from the left navigation menu on your WordPress dashboard, go to "Pages" and click "Add New."

After clicking the Add New button, the next section is same for both pages and posts.

In the first box, you will need to enter the title of your post or page.

Next box is known as the content editor box, where you actually write the content of your post or page. You will find two tabs on the top right side of this box, namely 'Visual" and "Text." By default, you will be provided with the Visual tab by WordPress. In the visual tab, you will actually see how your post will look like on your website, whereas if you switch to the Text tab, then you will see how your post is made using the text you provided and the codes that WordPress keeps inserting to maintain the post formatting. In the Text tab, you see the plain HTML version of your post. At the top right corner of your screen, you will see the "Publish" button. When you are done with writing your post or page, you will have to click this button to publish your content on the website. Here, you can save your post as a draft if you'd like to save it for later. If you click the "Preview" button, you can get a preview of how the post will look once it's published. The status of the post will show if the post has been published, saved as a draft, if it's pending review, or if it has been scheduled.

If you are working in posts, you will see a section called "Categories and Tags." These are the way to keep your content organized. You can create new categories for your website content from – "Posts" and then by clicking on "Categories," or you can do the same from the categories section on the "Add New" page of your post.

Now, let's discuss few important functions that you will use while creating your posts and pages. I am explaining these in the form of questions and answers to help you understand better. So, let's start.

How do I create links?

To create links in your posts and pages, just select the text that you want to link to, then click the hyperlink button from the toolbar present at the top of the content editor box. A new lightbox will now appear where you can add a new link for the selected text.

How do I add formatting?

Formatting is needed when you want to create a new heading or want to write something in a preformatted form. By default, in the top toolbar, you will see "Paragraph" as the default formatting set by WordPress for your post. You can select a lot of formatting options from the drop-down menu that appears when you click Paragraph.

How do I add images / other media?

In between the title box and content editor box, you will see a button called "Add Media." You can use this button to insert multimedia in your post. If you already have it in your gallery, then select one and click "Insert." Otherwise, go to upload tab and upload the file to your gallery first for inserting in the post.

These are the basic ideas behind how to create content using posts and pages in WordPress. You can go ahead and make an "About Us" page using this information on your WordPress blog.

How to Create a Contact Page?

In the plugins chapter of this eBook, I told you to install a plugin named "Contact Form 7." We will now use this plugin to create a contact page for your WordPress website. To create the page, you will first need to generate a contact form via Contact Form 7 plugin and then you will have to enter the short code of the form to a blank page which you want to use as a contact page. Below is a detailed tutorial for you to make a contact form within minutes.

- After installing Contact Form 7 and activating it, from the left navigation menu on your WordPress dashboard, go to "Contact" and click "Add New."

- On the next screen, enter the title of your contact form. Below the title box, you will see "Form," "Mail," "Messages," and "Additional Settings." Click on Mail.

- Now, enter your email in the "To" box. This is the email where you will receive all the messages which will be sent using the contact form.

- In the "From" box, write [your-name] <your email address here>

- In the "Subject," enter [your-subject]

- Now, leave everything else as it is. Scroll to the bottom of the page and click Save. You will be provided with a short code after you save the form. Copy that short code.

- Now, go to pages in the left navigation menu and click Add New.

- In the title box, write Contact Us.

- In the content editor box, write – "For any queries or suggestions, you can contact us using the contact form below. *(Paste the short code here in a new line that you copied before)."*

- Now, click "Publish" button and you are done!

You can create as many contact forms as you want using the above method. I hope this chapter helped you in understanding how you can create the content for your website.

Conclusion

I really hope that this eBook was helpful for you in creating a WordPress blog or website from scratch. But, this eBook is of no use if you will just read it and will keep it aside. Get online and put this eBook to use to gain maximum knowledge from this eBook.

CPSIA information can be obtained
at www.ICGtesting.com
Printed in the USA
LVHW041956151218
600583LV00001B/159/P